Order this book online at www.trafford.com
or email orders@trafford.com

Most Trafford titles are also available at major online book retailers.

© Copyright 2005, 2011 Kevin D Hunter.

All rights reserved. No part of this publication may be reproduced, stored in a retrieval system, or transmitted, in any form or by any means, electronic, mechanical, photocopying, recording, or otherwise, without the written prior permission of the author.

Printed in the United States of America.

ISBN: 978-1-4269-6476-3 (sc)
ISBN: 978-1-4269-6477-0 (e)

Library of Congress Control Number: 2011905739

Trafford rev. 04/11/2011

 www.trafford.com

North America & international
toll-free: 1 888 232 4444 (USA & Canada)
phone: 250 383 6864 ♦ fax: 812 355 4082

Introduction

Whether you are a high school student, attending college, going back to school or actively employed in the technical field, the need for knowledge of Trigonometry will likely pop up somewhere along the way. To many, the very mention of the word "Trig" sends chills down the spine and causes knots in the stomach. Society has long associated Trigonometry with the stereotype that only a genius, or those with a million years of college, can comprehend and utilize it. I disagree with those misconceptions. Given the proper method of teaching, almost anyone can learn it.

This book explains the basics of Trigonometry while simplifying all of the scientific jargon associated with so many other texts. It is written in such a way as to allow the average person to grasp trigonometric functions and Algebra without really having to know a whole lot about them. Only the necessary concepts are covered. The methods and narrative format are also unique because they are designed to break down and simplify, what may be otherwise, complicated information. There are a lot of illustrations and explanations to make things easier for people who do not posses an extensive knowledge of Mathematics or Geometry, especially those who are often confused or frustrated by conventional teaching methods.

A level of arithematic equal to or greater than "**Math II**", which is a common course in most middle schools, is the ideal comprehension that someone should achieve before attempting to learn Trigonometry. A fundamental knowledge of angles is also beneficial. However, enough is explained throughout the book that most individuals should be able to work through it and develop a good understanding of the information contained herein.

Do not try to become a master of Trigonometry overnight. Pace yourself as you go from chapter to chapter. Be rested and ready to focus during each session of learning and practice on the blank worksheets provided throughout the book as you progress through each topic. With a little time and patience, it will become clearly evident that expensive college courses or seminars are not necessary to learn this skill. Once the basics are covered, learning how to apply this knowledge will come with ease.

Sincerely

Kevin D Hunter
Author

Table of Contents

	Page #
Chapter 1	
The Law of the Mathematical Tripod	
Formula and Structure	1
Setting Up a Tripod	2
Chapter 2	
Understanding Triangles	
Identifying Right and Obtuse Triangles	3
Labeling Triangles for Trigonometry	4
Relationships between Sides and Angles	5
Chapter 3	
Angles and Their Rules	
180 Is the Magic Number	7
Inverse Angles	8
Chapter 4	
Using a Scientific Calculator for Trigonometric Functions	
Converting Angles to and from Sine, Cosine and Tangent	9
Understanding Squares and Roots	10
Chapter 5	
Solving Right Triangles	
The Three Tripods of Right Angle Trigonometry	11
The Pythagorean Theorem	15
Chapter 6	
Solving Obtuse Triangles	
The Law of Sines	17
The Law of Cosines	21
Chapter 7	
Miscellaneous Information	
Isosceles Triangles	29
Rules for Creating Your Own Practice Problems	30
Tips and Tricks	31
Quick Reference	33

The Not-So-Scary Guide to Basic Trigonometry

Written & Illustrated by
Kevin D Hunter

Chapter 1
The Law of the Mathematical Tripod

In order to simplify most Trig functions, one must understand and master the simple concept of the Mathematical Tripod. I call it a tripod because it always has three pieces of information. Using letters as **Variables** to represent numbers, a typical tripod is shown and its formulas defined below.

$$B = \frac{C}{A}$$ Definition:

$$A \times B = C$$
$$C \div B = A$$
$$C \div A = B$$

Don't let the formula intimidate you. It's basically a value equaling a fraction. A few simple rules will make this concept a no-brainer. Don't pay much attention to the = symbol right now either. It has its purpose in later chapters. Look at the multiplication string below.

4 × 5 = 20

To set up a tripod, always use the first formula, $A \times B = C$. Here is how it looks in a tripod.

$$5 = \frac{20}{4}$$

When setting up the formula, just remember to start on the bottom of the fraction and work **clockwise** until you get to the top. Think of it like going up a flight of stairs and then going back down. Start at the bottom on 4 and step on 5 to get up to 20. Then turn around and step on 5 to get back down to 4, and if you want to skip the stairs, just go straight from 20 down to 4. If you multiplied to get up the stairs, then you have to do the opposite and divide to get back down.

Try it out. Take **3 × 4 = 12** and fill in the tripod.

$$? = \frac{?}{?}$$ Answer: $4 = \frac{12}{3}$

Why is this so important? Using the same math from above, this time, something is left missing.

$$4 = \frac{12}{?}$$

Using simple logic, choose one of the three possible formulas from above that will correctly solve the missing piece of information. In this case, $C \div B = A$ is the correct choice.

12 ÷ 4 = ?

The answer is **3**

Here is a summary and breakdown from start to finish creating and defining a tripod using the multiplication problem below.

$3 \times 8 = 24$ Converted into a tripod: $8 = \dfrac{24}{3}$

Up the stairs: $3 \times 8 = 24$ $\longrightarrow$ With Variables: $A \times B = C$

Down the stairs: $24 \div 8 = 3$ $\longrightarrow$ With Variables: $C \div B = A$

Skip the stairs: $24 \div 3 = 8$ $\longrightarrow$ With Variables: $C \div A = B$

Practice Problems

Put the following multiplication strings into a tripod.

$6 \times 8 = 48$ = —   Answer: $8 = \dfrac{48}{6}$

$7 \times 4 = 28$ = —   Answer: $4 = \dfrac{28}{7}$

$3 \times 12 = 36$ = —   Answer: $12 = \dfrac{36}{3}$

Find the missing pieces of these tripods. The answers are written in a format that shows the operation that was performed to get the result.

$40 = \dfrac{120}{?}$ Answer: $120 \div 40 = \underline{3}$

$? = \dfrac{68}{10}$ Answer: $68 \div 10 = \underline{6.8}$

$7 = \dfrac{?}{9}$ Answer: $9 \times 7 = \underline{63}$

Using a tripod approach to solving Trig is simple and performs algebraic functions without using algebra the way it is conventionally taught in most classrooms. 100% of Right Angle Trigonometry and 33% of Obtuse Trigonometry can be easily solved using this method. Learning how a tripod works is your secret weapon. It is also what sets the teaching style of this book apart from so many others. Do not move on to the next chapter until the concepts learned from this one are thoroughly understood. You will soon be substituting parts of a triangle into special formulas based directly on the concept of a tripod.

FYI: Tripods aren't just useful for Trig. Try using them in other applications like Physics.
A good example is the formula: **Momentum = Mass $\times$ Acceleration**. In a tripod, it would look like this:

$$Acceleration = \dfrac{Momentum}{Mass}$$

If any two pieces of information were given, the third piece could be easily found by using this tripod.

WORKSHEET PAGES/NOTES

Chapter 2
Understanding Triangles

To successfully master Trigonometry, a basic understanding of triangles is necessary. In this chapter, you will learn how to identify the two basic types of triangles, name the different parts, understand the relationships between those parts and label them for solving in trigonometric tripods that will be introduced later in the book.

Anatomy of a Right Triangle

The most common type of triangle is the Right Triangle. Any triangle with a 90-degree corner is a Right Triangle because the 90-degree angle is known in Geometry as a "**Right Angle**". A Right Triangle looks like this:

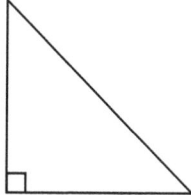

These are the easiest to solve and are commonly labeled in diagrams and drawings with a small box in the corner where the right angle is located. Every triangle has three sides and three angles. The sides of a Right Triangle are defined as follows:

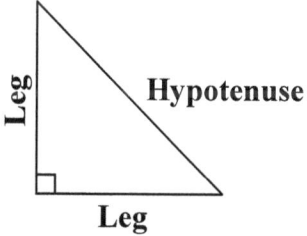

It is important to learn how to find and identify the parts of a triangle correctly. Do not let the word Hypotenuse intimidate you. The odd name will actually help with memorization of the necessary formulas later. The two sides of a Right Triangle that meet to form the 90-degree corner are always known as Legs, and the third side that connects them is known as the Hypotenuse. It is also useful to know that the Hypotenuse is always the longest side and is directly across from the 90-degree angle.

Label the sides of the triangles below.

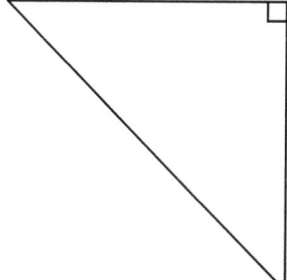 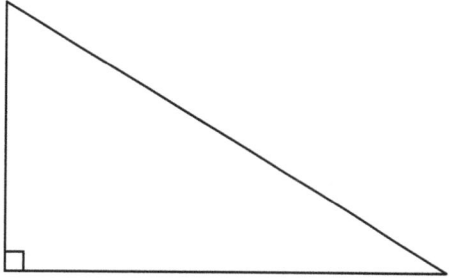

3

Obtuse Triangles

Obtuse Triangles are defined as any type of triangle that does not contain a 90-degree angle. They can also be a lot harder to solve than Right Triangles. A typical Obtuse Triangle looks something like this:

Obtuse Triangles always have two acute angles and a third angle that can be either acute or obtuse, depending on the dimensions. An **Acute Angle** is any angle less than 90 degrees but greater than zero. An **Obtuse Angle** is any angle greater than 90 degrees but less than 180. These triangles also have no **Legs** or a **Hypotenuse**.

Labeling Triangles for Trigonometry

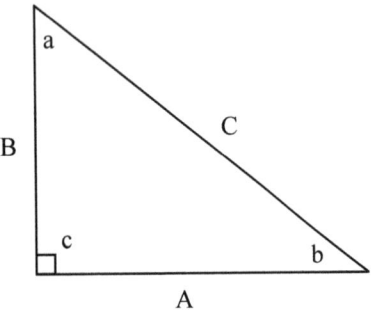

Take a close look at the labeling of the triangle above. Throughout this book, the uppercase and lowercase letters **A, B, C & a, b, c** will be used to label the different parts of a triangle for solving as **Variables**. The uppercase letters will represent **Sides** and **Legs**, while the lowercase letters will represent **Angles**.

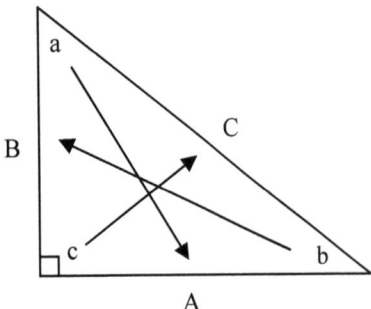

The formulas in later chapters will be a lot easier to understand if the triangles are labeled like the one above. Write in all of the angles first with lowercase letters, then label the sides of the triangle **directly opposite** from their angles with the matching uppercase letters. Always label the 90-degree corner as **Angle c** and the **Hypotenuse** as **Side C**. If you already have a labeling convention that works, use that instead.

Relationships between Sides and Angles

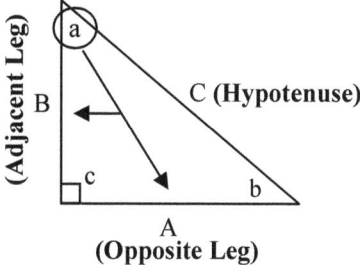

Above is a Right Triangle with correct labeling. There exists a relationship between the angles and sides that is vital to understand when solving Trig formulas. To give an example, take a look at the angle labeled with the lowercase "a". The **Opposite Leg** is obviously **Side A**, but what does the **Adjacent** label mean? The **Leg** of the triangle that is immediately <u>next to the angle</u> is known as the **Adjacent Leg** from that angle.

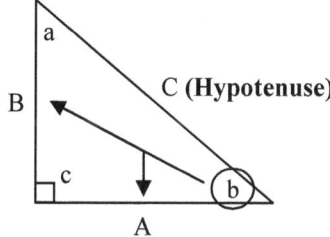

In the triangle above, which **Leg** is **Adjacent** from **Angle b**? Which **Leg** is **Opposite**?

<u>Answers:</u> **Adjacent Leg = Side A**
Opposite Leg = Side B

Obtuse Triangles are labeled the same as Right Triangles with the exception that there is no **Hypotenuse**. To make things easier while learning from this book, if there is an Obtuse Triangle with an angle greater than 90 degrees, it should be labeled as **Angle c**. The other two angles should be labeled **Angle a** and **Angle b** respectively. An Obtuse Triangle with proper labeling looks like this:

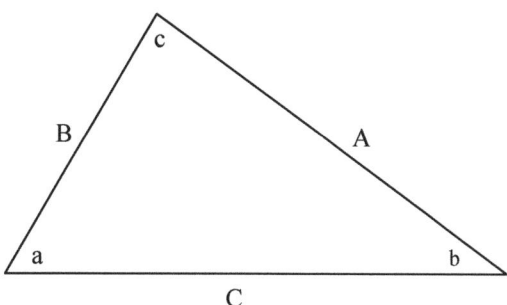

NOTE: These are a breed of their own and also require more difficult math to solve than traditional Right Triangles. There are no **Legs** or **Hypotenuse**; only **Angles** and their **Opposite Sides**.

5

Practice Problems

Identify and label the triangles below using letters as Variables.

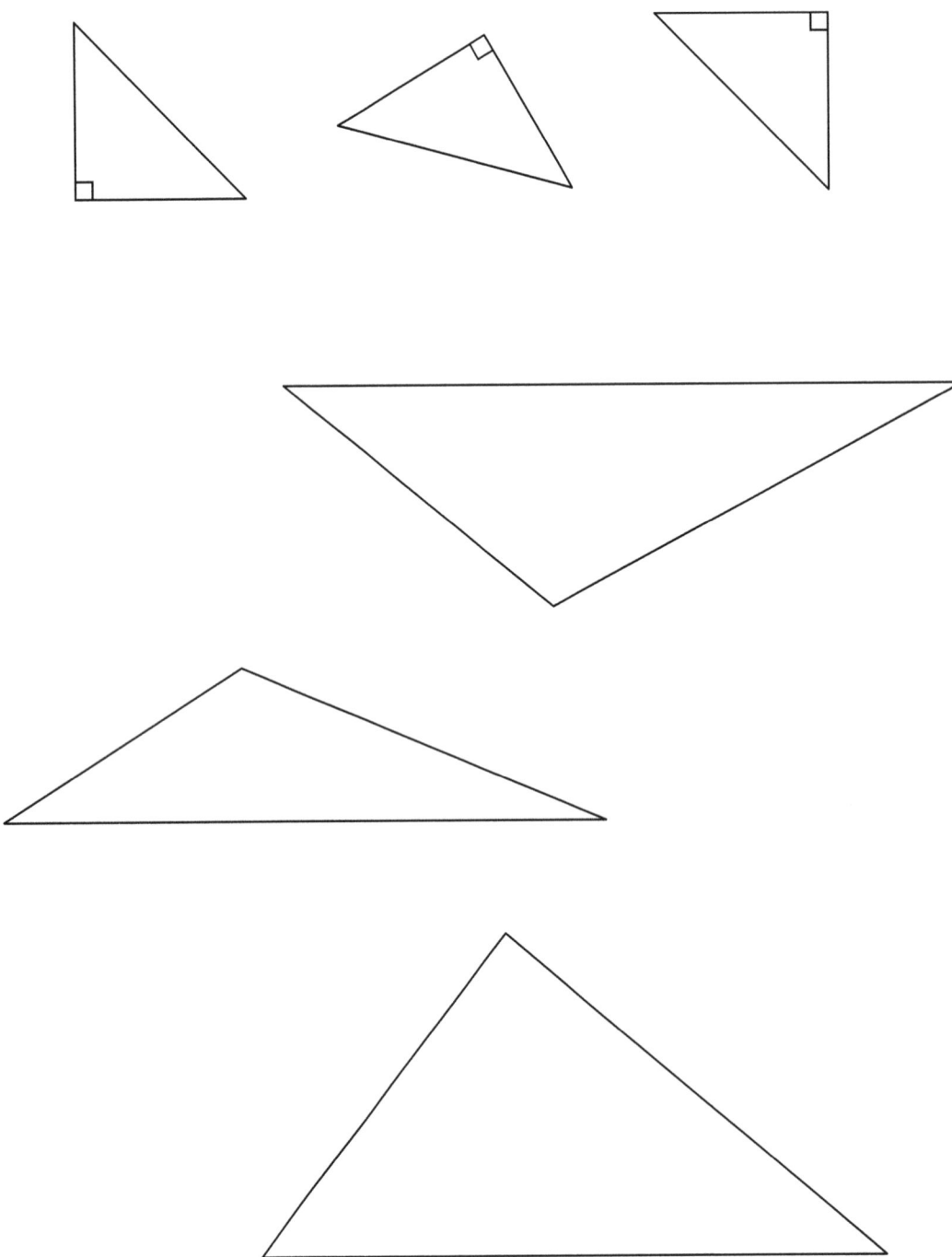

WORKSHEET PAGES/NOTES

Chapter 3
Angles and Their Rules

180 Is the Magic Number

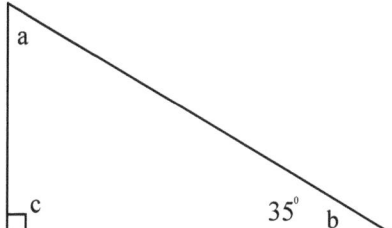

There is one very important rule governing angles in Trigonometry. The rule states: **All three angles, when added together, must equal 180 degrees**. This simple fact will be invaluable for solving angles later on.

Take the triangle above, for instance. In Right Triangles, we always know that the corner with the square box is a 90-degree angle. We can also deduct that 90 is ½ of 180, so the other two angles must always add up to 90 degrees in order for the triangle to be true. **Angle b** is 35 degrees, so **Angle a**, when added to **Angle b**, must add up to 90 degrees. A simple subtraction problem should do the trick.

90 – 35 = 55

So, the missing angle is **55 degrees** because **55 + 35 = 90**. In Obtuse Trig, there is no 90-degree angle. There will need to be two of the three angles given in order to solve the third, which can be found by subtracting the other two angles from 180.

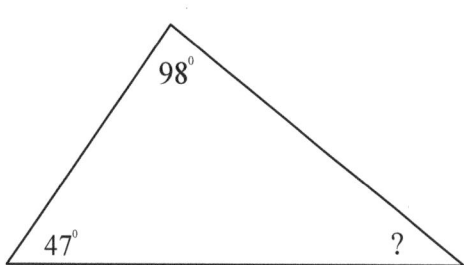

In this triangle, use the given information to find the missing angle. Remember that all of the angles must add up to equal 180.

Answer:

180 – 98 – 47 = 35

Our missing angle is **35 degrees** (**98 + 47 + 35 = 180 degrees**)
All three angles are now defined.

7

Inverse Angles

During Obtuse Trigonometry, there will often be angles greater than 90 degrees. Although a calculator will give a correct answer when converting a decimal back into an angle greater than 90 degrees using **Cosine**, it may be necessary to use the **Inverse** of an angle to end up with the right answer when a decimal has been converted back into an angle using **Sine** or **Tangent**.

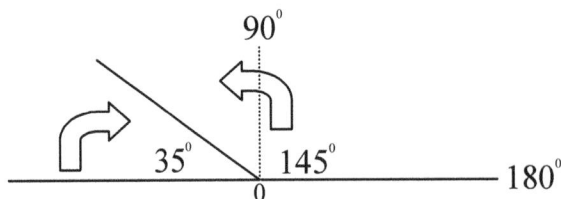

An angle's inverse can be found simply by subtracting it from 180. In the example above, the inverse of 145 degrees is 35.

180 – 145 = 35

Find the inverse of these angles:

$115°$ Answer: **65 degrees**

$98°$ Answer: **82 degrees**

$166°$ Answer: **14 degrees**

$63°$ Answer: **117 degrees**

Do not worry too much about this until chapter 6 when we start solving Obtuse Triangles. There are instances where an **Inverse Angle** must be calculated and used to get correct information for an angle while using the **Law of Sines** much later in the book.

Chapter 4
Using a Scientific Calculator for Trigonometric Functions

If I were to go into a mathematical war zone, a scientific calculator would be my M-16. If you are already familiar with your calculator, simply move on to chapter 5. Make sure it is in **Degree Mode** before using it for any Trig functions. The letters **DEG** should be visible just above the zero. If it is not present anywhere on the display, consult the users' manual to find out how to get into **Degree Mode**. Most scientific calculators enter this mode by default whenever they are turned on.

Converting Angles to and from Sine, Cosine and Tangent

Using a calculator to convert angles is a relatively simple process. This skill is important because **an angle must always be converted to its decimal SINE, COSINE or TANGENT before it can be used in any trigonometric equation**. Visa versa, when solving for angles, **a decimal must always be converted back into an angle in order to get the correct dimension**.

Take a 30-degree angle as an example. Start out by using the **Sine** function first. To convert 30 degrees to its decimal in **Sine**, press **30** and then hit the **SIN** button on the calculator. You should see **0.5** on the display. Now try the **COS** and **TAN** buttons using the same angle and repeat the steps above.

30° SINE = 0.5

30° COSINE = 0.866025404

30° TANGENT = 0.577350269

Try this with other angles to get a feel for it. Now that you are familiar with converting angles to their respective decimals, let's move on to learning how to convert a decimal back into an angle. Take the 30-degree angle again. We know that the **Sine of 30 degrees is equal to 0.5**, so let's convert it back to an angle. Depending on the type of calculator being used, the locations and text of the keys may vary. However, on most scientific calculators, there is usually a smaller print of text just above the **SIN, COS** and **TAN** buttons labeled **SIN^{-1}, COS^{-1}** and **TAN^{-1}**. Using the **Shift** or **2nd** key, and then pressing the corresponding **SIN, COS** or **TAN** key, will convert the decimal back into its original angle. If there is difficulty finding the functions on the calculator, consult the owners' manual.

Type **0.5** into the calculator. Now press the **Shift** or **2nd** key. Then press the **SIN** key. You should see **30** in the display.

Practice Problems

Convert these decimals into angles using the designated keys.

Use **SIN^{-1}**   ⇒   0.906307787

Use **COS^{-1}**   ⇒   0.422618262

Use **TAN^{-1}**   ⇒   2.14450692

The answer to all three conversions is **65 degrees** when rounded to the nearest whole number.

Understanding Squares and Roots

Understanding how to calculate squares and roots of numbers is vital to learning and solving Trig formulas. Squaring a number is a relatively simple concept. Just take any number and multiply it by itself, and you have **squared** it.

The square of <u>4</u>: $4 \times 4 = 16$ is written as: $4^2 = 16$

The square of <u>9</u>: $9 \times 9 = 81$ is written as: $9^2 = 81$

On the calculator, there is a button with the x^2 symbol written on it. This is the button used to square a number. Also, located right next to or nearby the x^2 key, is the $\sqrt{x}$ key. This is known as the **Root** key. Pressing that one will reverse the square of a number and give the original number that was squared.

Example:
Press **10** and then the x^2 key on the calculator. The result is **100**. Now, hit the $\sqrt{x}$ key. The result is **10** because the square was reversed to give the original figure.

$10^2 = 100$

$\sqrt{100} = 10$

Practice Problems

Square the following numbers. Then practice using the **Root** key to reverse the square.

7^2 Answer: **49**

5^2 Answer: **25**

8^2 Answer: **64**

Now try adding squares together using a Variable for the result.
Hint: The keys are pressed on the calculator for the first problem like this: **4** ⇨ x^2 ⇨ + ⇨ **6** ⇨ x^2 ⇨ =

$4^2 + 6^2 = A$ Answer: **A = 52**

$5^2 + 7^2 = B$ Answer: **B = 74**

$2^2 + 3^2 = C$ Answer: **C = 13**

You now have all of the basics necessary to begin learning how to solve Right Triangles in the next chapter!

WORKSHEET PAGES/NOTES

Chapter 5
Solving Right Triangles

The Three Tripods of Right Angle Trigonometry

Listed below are the three tripod formulas for solving information in Right Triangles. Although it is not required, I strongly recommend that these be memorized or made readily available for reference while learning from this book. Spend a little time figuring out a memory scheme that will stick. If these can be learned by heart, difficult triangles can be solved very quickly by visualizing the formulas in your head and then typing the multiplication or division into the calculator according to the laws of the tripod instead of writing the information down.

Sine Angle = $\dfrac{\text{Opposite Leg}}{\text{Hypotenuse}}$

Cosine Angle = $\dfrac{\text{Adjacent Leg}}{\text{Hypotenuse}}$

Tangent Angle = $\dfrac{\text{Opposite Leg}}{\text{Adjacent Leg}}$

These simple tripods are the keys to 75% of all Trigonometry. **It is important to remember, however, that a tripod can only be solved if two of the three pieces are given.** Another fact that needs to be taken into consideration is that a triangle cannot be solved if there is not at least one **Leg** or **Side** provided. This is because the **Legs** of the triangle essentially define its finite properties including respective angles in relation to other **Legs** or **Sides**. Now, that said, all of the hard work from chapter 1 is about to pay off.

The first skill to develop is deciding which formula to use for a particular situation in a Right Triangle. Take a look at the one below.

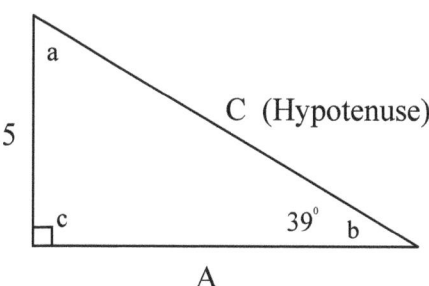

First, check that the triangle can be solved. There are two pieces given, and one is a **Leg**, so this one can be solved. Now choose which formula should be used to find the rest of the information for **Angle a, Side A** and **Side C**, which is the **Hypotenuse**.

NOTE: The triangles depicted throughout this book are not drawn to scale. Therefore, the dimensions that are solved may not accurately depict how the triangle looks with the naked eye. A protractor and ruler cannot be used to cheat. ☺

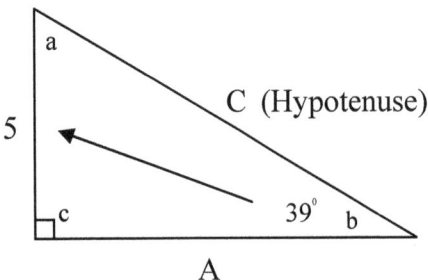

Analyze the information that has been given. There is an **Angle** and its **Opposite Leg**. If we refer to the formula chart, we can see that there are two formulas that contain an **Angle** and an **Opposite Leg**.

Sine   Angle = $\dfrac{\text{Opposite Leg}}{\text{Hypotenuse}}$

Tangent   Angle = $\dfrac{\text{Opposite Leg}}{\text{Adjacent Leg}}$

By looking closely at the formulas, it can be analyzed, if the given information is plugged in, that the **Hypotenuse** can be found by using the formula for **Sine**, and the **Adjacent Leg**, which is **Side A**, can also be found by using the formula for **Tangent**. So, depending on which dimension is needed, we would choose one or both formulas to find what we are trying to solve. **Remember, before any angle can be used in a trigonometric function, it must first be converted to a decimal.**

Fill in the first formula with the dimensions that are given in the triangle.

39 SIN = $\dfrac{5}{\text{Hypotenuse}}$ Now rewrite the formula for calculation using the law of the tripod.

5 ÷ 39 SIN = Hypotenuse   On your calculator, type: **5 ⇨ ÷ ⇨ 39 ⇨ SIN ⇨ =**

Result: **Hypotenuse (Side C) = 7.945078645**

Now set up the tripod for the next formula.

39 TAN = $\dfrac{5}{\text{Adjacent Leg}}$ Now rewrite the formula for calculation

5 ÷ 39 TAN = Adjacent Leg (Side A)

Result: **Adjacent Leg (Side A) = 6.174485783**

All of the dimensions are now solved except for **Angle a**. Remember the rules of angles covered in chapter 3? **Angle a** and **Angle b** must add up to equal 90 degrees. So find **Angle a** by subtracting the 39-degree angle from 90.

90 – 39 = 51
Angle a = 51 degrees

In the first triangle we solved, there was an angle defined that was used to find the other angle, but what if there were only two sides of a triangle given? How would one of the angles be found in this case?

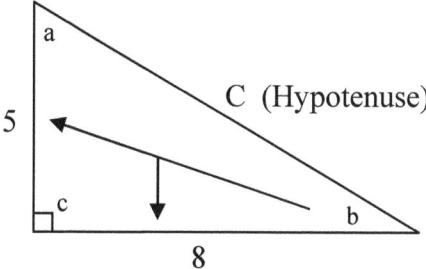

Examine the triangle to see what is given. That's right. Two **Legs** are provided. Everything else is a mystery. This triangle is quite different. If either **Angle a** or **Angle b** can be found, subtraction from 90 degrees can be used to find the other one. Which angle should be solved? Both have an **Opposite Leg** and an **Adjacent Leg** defined, so in this case, it doesn't really matter which one is used. For this example, use **Angle b**.

Look to the formula chart to see which one has an **Opposite Leg** and an **Adjacent Leg** in it.

Tangent   Angle = $\dfrac{\text{Opposite Leg}}{\text{Adjacent Leg}}$

The **Tangent** formula is the only one that has both of the things that were given, so fill in the formula with the information we already have.

Angle b TAN^{-1} = $\dfrac{5}{8}$ Rewrite the formula for calculation.

5 ÷ 8 = The Decimal Tangent of Angle b

When 5 is divided by 8, we end up with the **Decimal Tangent** of **Angle b**. Don't forget, when solving to find angles, **the decimal must always be converted back into an angle** to get the correct answer. If you are a bit rusty on remembering how to do this, refer back to chapter 4.

Angle b TAN^{-1} = 0.625   Convert **0.625** back into an angle

Angle b = 32.00538321

Now find **angle a** by subtracting **Angle b** from 90 degrees.

90 − 32.00538321 = Angle a

Angle a = 57.99461679

13

Practice Problems

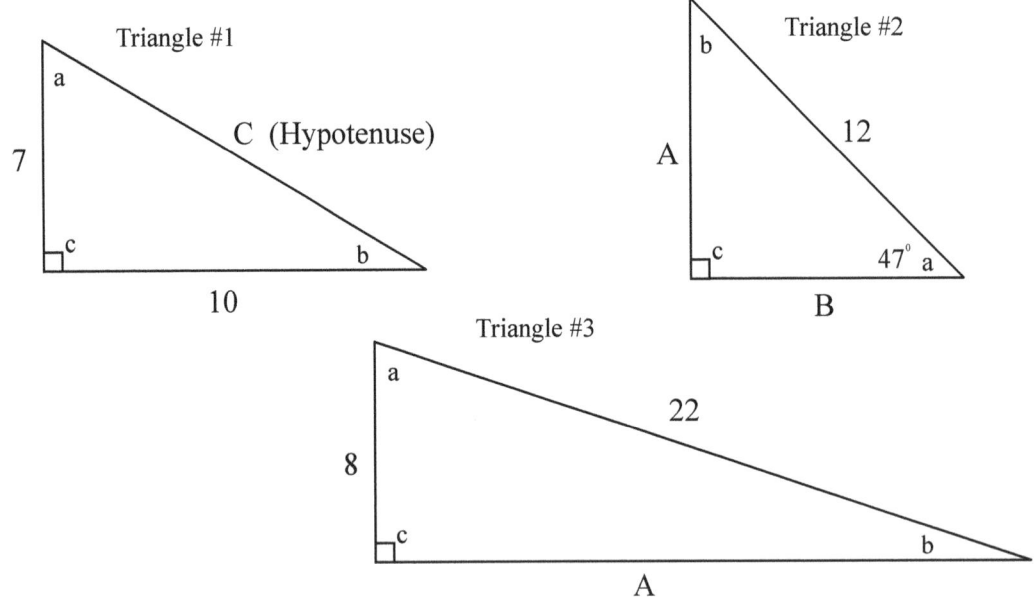

Use the information provided in each triangle to find the other missing sides and angles. Sometimes one formula must be solved in order to get information needed for another formula to find a dimension. Triangle #1 is a good example. To find the **Hypotenuse**, a **Tangent** formula with the **Legs** will have to be solved to get an angle first. Then either **Sine** or **Cosine** can be used with the angle to get the **Hypotenuse**. The only difference between **Sine** and **Cosine** is that they use different **Legs** of the triangle in relation to the angle to get the **Hypotenuse**. Since both **Legs** of the triangle are given, it doesn't matter which of the two formulas is used after an angle is found.

Answers:

Triangle #1

Angle a = 55.0079798
Angle b = 34.9920202
Hypotenuse = 12.20655562

Triangle #2

Angle b = 43
Side A = 8.776244419
Side B = 8.183980321

Triangle #3

Angle a = 68.67631374
Angle b = 21.32368626
Side A = 20.49390153

The Pythagorean Theorem

Now that you are comfortable using **Sine**, **Cosine** and **Tangent** to find all of the different parts of a Right Triangle, let's examine a popular shortcut. There exists a wonderful mathematical concept known as the **Pythagorean Theorem**. This theorem only applies when looking for a missing side of a Right Triangle when the other two sides are provided. **No angles are involved!**

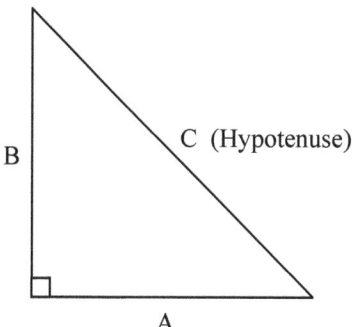

In the triangle above, the **Pythagorean Theorem** states:

$A^2 + B^2 = C^2$
$C^2 - B^2 = A^2$
$C^2 - A^2 = B^2$

Using a scientific calculator, it is easy to quickly find the missing leg. Don't forget that **Side C** is **always** the **Hypotenuse**. It will be easy to remember what to do if this is thought about logically rather than technically with the formulas. There are only two combinations or scenarios that exist with this theorem. Either the **Hypotenuse** and one of the **Legs** are given, or both **Legs** are provided. If both **Legs** are provided, just add the squares of the **Legs** and then use the **Root** key to get the **Hypotenuse**. If one of the **Legs** and the **Hypotenuse** are given, then subtract the square of the **Leg** from the square of the **Hypotenuse** and then use the **Root** key to find the other **Leg**. This theorem is very easy to use and is quite effective as a shortcut when two sides are available and the third one is needed quickly.

Look at the triangle below.

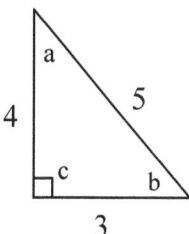

The combinations of formulas in step-by-step analysis are defined:

$3^2 + 4^2 = 5^2$ ⟹ $9 + 16 = 25$ ⟹ Hypotenuse = $\sqrt{25}$
$5^2 - 4^2 = 3^2$ ⟹ $25 - 16 = 9$ ⟹ Side A = $\sqrt{9}$
$5^2 - 3^2 = 4^2$ ⟹ $25 - 9 = 16$ ⟹ Side B = $\sqrt{16}$

Practice Problems

Solve the missing pieces of these triangles using the **Pythagorean Theorem**.

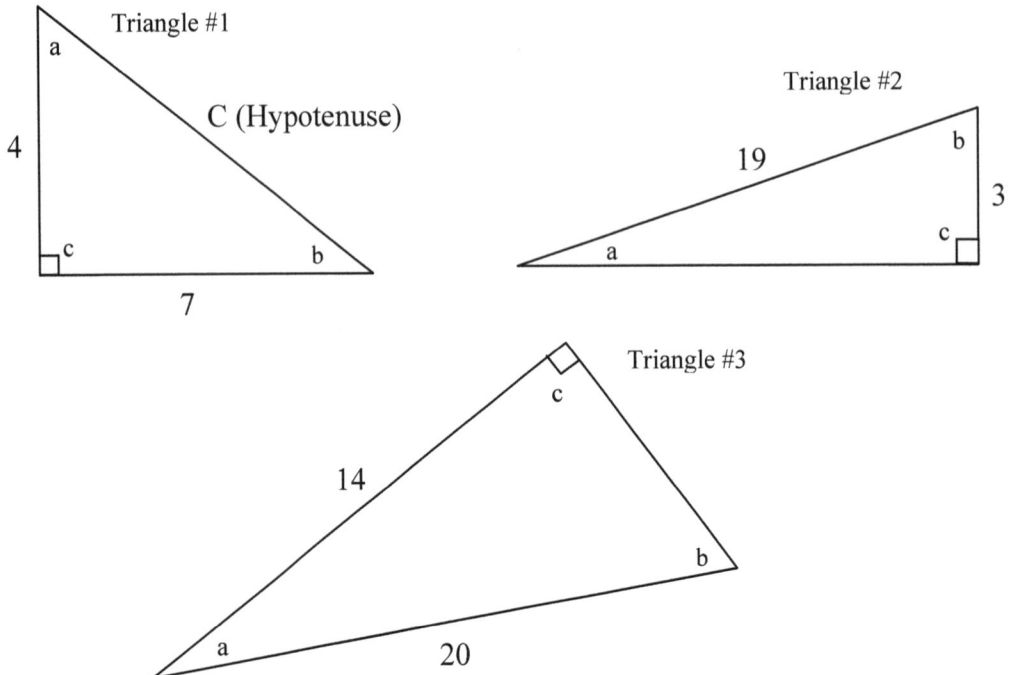

Answers given in step-by-step format:

Triangle #1

$7^2 + 4^2 =$ **Side C** ^{2}
$49 + 16 = \underline{65}$
Side C $= \sqrt{65}$
Side C = 8.062257748

Triangle #2

$19^2 - 3^2 =$ **Side B** ^{2}
$361 - 9 = \underline{352}$
Side B $= \sqrt{352}$
Side B = 18.76166304

Triangle #3

$20^2 - 14^2 =$ **Side A** ^{2}
$400 - 196 = \underline{204}$
Side A $= \sqrt{204}$
Side A = 14.28285686

WORKSHEET PAGES/NOTES

Chapter 6
Solving Obtuse Triangles

We learned in chapter 2 how Obtuse Triangles are quite different from Right Triangles. There is no 90-degree angle and no Hypotenuse to simplify things. Be alert and ready to focus during this chapter. There are essentially three different types of Obtuse Triangles, each with their own method for solving. The first type utilizes the **Law of Sines** to find the missing pieces. There is one important detail that should be noted. **There must be three given pieces of information in all Obtuse Triangles in order to solve them**, and like Right Triangles, **one of those pieces must be a Side**. If this information is not available, the triangle cannot be completed.

The Law of Sines

This law applies when: **An ANGLE and its OPPOSITE SIDE and any other dimensions are given**.

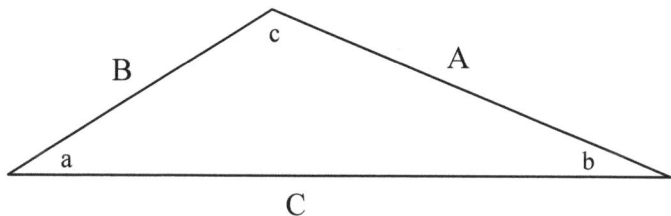

The law reads:

Side A ÷ the Sine of Angle a is the same as ⇩
Side B ÷ the Sine of Angle b, which is also the same as ⇩
Side C ÷ the Sine of Angle c

For you Algebra junkies, it's written like this:
(A/SIN a) = (B/SIN b) = (C/SIN c)

Each **Side** of the triangle, when divided by the **Sine** of its **Opposite Angle**, gives the same result. This unique property allows for the determination of a **Constant**, or **Key**, that can be used to find the missing pieces. Fortunately for us, we can still use our trusty tripod method to compute the **Law of Sines**. The three tripods for finding the **Key** in the triangle above are defined below.

$$\text{Key} = \frac{\text{Side A}}{\text{The Sine of Angle a}}$$

$$\text{Key} = \frac{\text{Side B}}{\text{The Sine of Angle b}}$$

$$\text{Key} = \frac{\text{Side C}}{\text{The Sine of Angle c}}$$

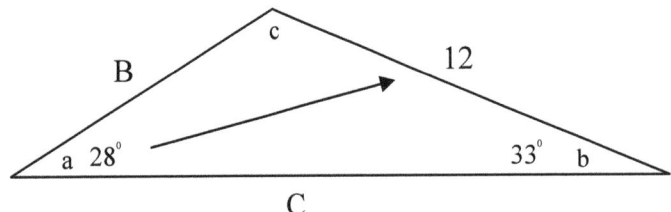

In the triangle above, **Angle a** and **Side A** are a pair of given dimensions that allow for the use of the **Law of Sines** to find a **Key**. Make a tripod using this law to find the **Key**.

$$\text{Key} = \frac{12}{28 \text{ SIN}}$$ Converts to: **12 ÷ 28 SIN = Key**   So the Key = 25.56065362

Remember, the angle must be converted to its **Sine** before the = button is pushed on the calculator. This is the order in which they should be pressed:

12 ⇨ ÷ ⇨ 28 ⇨ SIN ⇨ =

If the calculator has a storage function, this would be a good time to use it. Store the **Key** into the memory so it can be recalled while solving the other tripods. Use the **Key** to solve another piece of missing information. **Angle b** is given, so start there. Set up the tripod using **Angle b** and the **Key** we just found and use it to find **Side B**.

25.56065362 (Key) = $\dfrac{\text{Side B}}{33 \text{ SIN}}$ Converts to: **33 SIN × 25.56065362 = Side B**

Side B = 13.92132972

Side A, Angle a, Side B and **Angle b** are now defined, but what about **Side C** and **Angle c**? Neither dimension is given. Remember the rules about angles in triangles. All angles must add up to equal 180. Since two of the three angles are defined, just subtract them from 180 to find the final angle. Once the angle is solved, it can be used to find the final dimension for **Side C**.

180 – 28 – 33 = Angle c

Angle c = 119 degrees

Now set up a tripod to find **Side C**

25.56065362 (Key) = $\dfrac{\text{Side C}}{119 \text{ SIN}}$ Converts to: **119 SIN × 25.56065362 = Side C**

Side C = 22.35585138

The triangle is now fully solved. Remember to use the **Law of Sines** whenever possible because it is the most simple way to solve Obtuse Triangles.

Practice Problems

Use the **Law of Sines** to solve the missing dimensions of the triangles below.

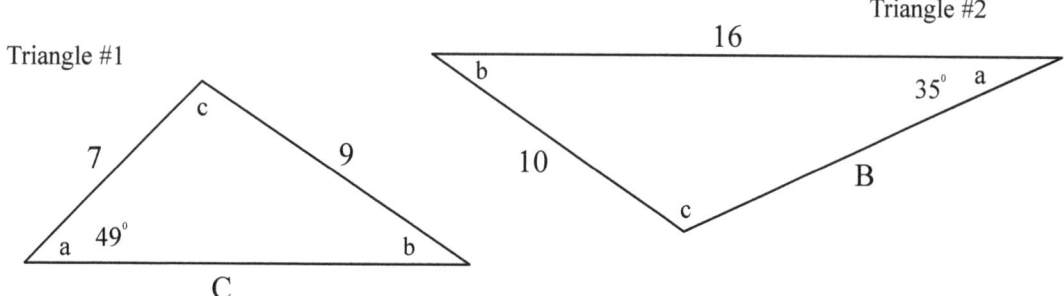

Answers:

Triangle #1

Key = $\dfrac{9}{49 \text{ SIN}}$ Converts to: **9 ÷ 49 SIN = Key**   So the Key = **11.92511694**

Now find **Angle b**

11.92511694 (Key) = $\dfrac{7}{\text{SIN b}}$ Converts to: **7 ÷ 11.92511694 = The Sine of Angle b**

Convert the decimal back into an angle.

0.58699634 = The Sine of Angle b
Angle b = 35.94414785

Subtract the two angles to find **Angle c**.

180 – 49 – 35.94414785 = Angle c
Angle c = 95.05585215

Now that **Angle c** is solved, make the last tripod to find **Side C**.

11.92511694 (Key) = $\dfrac{\text{Side C}}{95.05585215 \text{ SIN}}$

Converts to: **95.05585215 SIN × 11.92511694 = Side C**

Side C = 11.87871947

The triangle is solved.

19

Triangle #2

$$\text{Key} = \frac{10}{35\,\text{SIN}}$$ Converts to: **10 ÷ 35 SIN = Key**   So the Key = **17.43446796**

Now find **Angle c**

17.43446796 (Key) $= \dfrac{16}{\text{SIN c}}$ Converts to: **16 ÷ 17.43446796 = The Sine of Angle c**

Convert the decimal back into an angle.

0.917722298 = The Sine of Angle c
Angle c = 66.59533661 ???
Wait a minute!!! What's wrong with this answer???

By looking at the triangle, it is obvious that **Angle c** is greater than 90 degrees. The **Law of Sines** has one nasty drawback. It will never allow a decimal to be converted back into an angle greater than 90 degrees. This is why **Angle c** should always be labeled across from the longest side, because the angle across from the longest side of any Obtuse Triangle is the one that will be larger than 90 degrees if the situation exists. Remember what was written about **Inverse Angles** in chapter 3? This concept must be applied in order to find the right dimension for **Angle c**.

Convert the result for **Angle c** to its **Inverse Angle**

180 – 66.5953361 = The true result for Angle c
Angle c = 113.4046634

Subtract the two angles to find **Angle b**.

180 – 35 – 113.4046634 = Angle b
Angle b = 31.59533661

Now that **Angle b** is solved, make the last tripod to find **Side B**.

17.43446796 (Key) $= \dfrac{\text{Side B}}{31.59533661\,\text{SIN}}$

Converts to: **31.59533661 SIN × 17.43446796 = Side B**

Side B = 9.134206841

The triangle is now complete.

Note: Remember that these triangles are not drawn to scale. Even though **Side B** and **Angle b** look larger than **Side A** and **Angle a** in the drawing, the math tells a different story. Learning to trust the arithmetic and not the eye is a valuable skill that will be useful in the real-life application of Trigonometry. The only exception is **Angle c**. This angle will always be drawn wider to avoid confusion when dealing with **Inverse Angles**. To avoid this problem altogether, simply solve for **Angle a** and **Angle b** first and then subtract them from 180 whenever possible.

WORKSHEET PAGES/NOTES

The Law of Cosines

The **Law of Sines** is good when solving triangles where an **Angle** and its **Opposite Side** and any other dimension are given, but what if we encoutered a triangle like the one below?

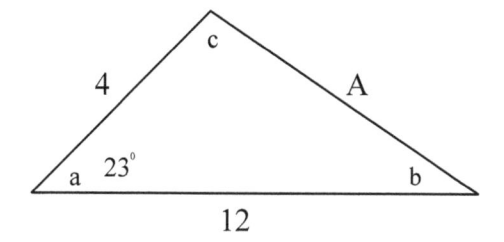

```
         c
    4  /  \   A
      /    \
   a/ 23°   \b
   ----------
       12
```

There are three pieces of information, and one is a **Side**, so it can be solved, but the **Law of Sines** does not work here because there is no **Angle** and its **Opposite Side** provided. A key cannot be found to unlock the rest of this triangle, so what do we do? This situation calls for the **Law of Cosines**.

The law is defined:

To solve for **Side A** or **Angle a**
Side A^2 = **Side B**^{2} + **Side C**^{2} - **2** × **Side B** × **Side C** × **The Cosine of Angle a**

To solve for **Side B** or **Angle b**
Side B^2 = **Side A**^{2} + **Side C**^{2} - **2** × **Side A** × **Side C** × **The Cosine of Angle b**

To solve for **Side C** or **Angle c**
Side C^2 = **Side A**^{2} + **Side B**^{2} - **2** × **Side A** × **Side B** × **The Cosine of Angle c**

Do not let these algebraic formulas scare you. They are not very difficult to understand when broken down and simplified. Again, you will be computing algebra without even knowing it. Coupled with the knowledge of the tripod formulas, the rules for this type of equation are all that are necessary to master Trigonometry.

Once again, upon examination, it is obvious that we are dealing with an **Angle** and its **Opposite Side**. The **Side** is always at the far left of the equation, and the **Angle** is always at the far right. Knowing this will make it easier to choose which formula to use when solving for missing information.

In the triangle above, **Angle a** is given, so if **Side A** can be found, the **Law of Sines** can be used to find a **Key** and solve the rest of the triangle. The **Law of Cosines** will allow for the computation of **Side A** using the information provided. This is the most efficient way to solve a triangle like the one shown on this page.

Use the formula from the previous page that allows for the computation of **Side A** as an example. The formulas can be broken down into three parts.

| Problem #1 | Problem #2 | | Problem #3 |

$$[\text{Side A}^2] = [\text{Side B}^2 + \text{Side C}^2] \pm [-2 \times \text{Side B} \times \text{Side C} \times \text{The Cosine of Angle a}]$$

Think of this formula as three miniature math problems all crammed into one long mess. There is a series of steps that need to be taken in order to come up with a single answer. There are also a couple of things that need to be clarified about the formula itself. Look at the arrow pointing to the **-2** and the other arrow pointing at the **plus-or-minus** symbol. The **-2** is a **Constant**. This means that a **-2** will always be in every formula and will not be a different number at any time for any reason. It's just there, and it has a special purpose. That's all you need to know.

The **plus-or-minus** symbol is placed there only to help as a visual reminder to either add or subtract the results of **Problem #3** from the solution of **Problem #2**, depending on whether the result from **Problem #3** was a positive or negative number. If the number is a positive number, they are added together. If the result is a negative number, then **Problem #3** is subtracted from **Problem #2**.

That said, the first step to completing the formula is to fill in the information from the triangle.

| Problem #1 | Problem #2 | | Problem #3 |

$$[\text{Side A}^2] = [4^2 + 12^2] \pm [-2 \times 4 \times 12 \times \text{The Cosine of 23 degrees}]$$

Complete the math in the brackets. **Problem #3** may present a challenge if you are not sure how to enter the information into the calculator. If there is not a **positive-to-negative conversion button,** simply type the subtraction or minus symbol **before** the number **2** is pressed. If it is done that way, however, a negative number will not be displayed at any time during the calculation until the = key is pushed at the end. **Problem #3** is typed into the calculator like this:

-2 ⇨ × ⇨ 4 ⇨ × ⇨ 12 ⇨ × ⇨ 23 ⇨ **COS** ⇨ =

Rewrite the formula with the solutions from all of the brackets:

Since the result from **Problem #3** is a negative number, Set up the next solution as a subtraction problem and combine the results from **Problem #2** and **Problem #3** into a single set of brackets.

$$[\text{Side A}^2] = [160 - 88.36846593]$$

Complete the subtraction in the second set of brackets. Then rewrite the formula:

Side A^2 = 71.63153407

Use the square root to find **Side A**.

Side A = $\sqrt{71.63153407}$

Side A = 8.463541461

Now that **Side A** is solved, use the **Law of Sines** to finish the rest of the triangle.

Practice Problems

Use the **Law of Cosines** to find the missing **Side**, then use the **Law of Sines** to find all of the other missing information in both triangles.

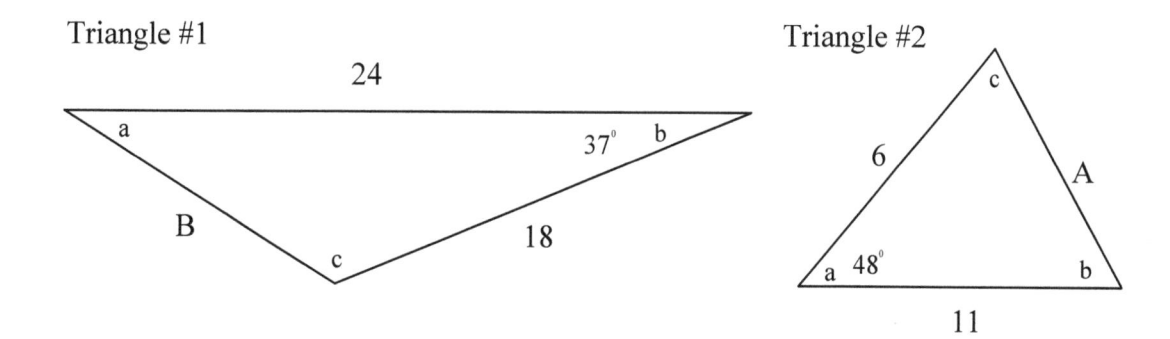

Triangle #1

Triangle #2

Answers:

Triangle #1

Use the formula to solve for **Side B** and fill in the information from the triangle.

Side B^2 = Side A^2 + Side C^2 - 2 × Side A × Side C × The Cosine of Angle b

Side B^2 = 18^2 + 24^2 - 2 × 18 × 24 × The Cosine of 37 degrees

If you need to use brackets and divide the problem up into three parts in order to avoid confusion, do so at this time. Then complete the solution for each set of brackets and rewrite the problem.

Side B^2 = 900 − 690.0210807

Complete the subtraction.

Side B^2 = 209.9789193

Use the square root of the solution to find the dimension for **Side B**

Side B = $\sqrt{209.9789193}$

Side B = 14.49064938

Hint: While solving the rest of the triangle, find **Angle a** first in case **Angle c** is greater than 90 degrees.

Triangle #2

Use the formula to solve for **Side A** and fill in the information from the triangle.

Side A ^{2} = Side B ^{2} + Side C ^{2} - 2 × Side B × Side C × The Cosine of Angle a

Side A ^{2} = 6^2 + 11^2 - 2 × 6 × 11 × The Cosine of 48 degrees

Solve the solutions from the problems and rewrite the formula.

Side A ^{2} = 157 − 88.32524004

Complete the subtraction.

Side A ^{2} = 68.67475996

Use the square root of the solution to find the dimension for **Side A**

Side A = $\sqrt{68.67475996}$

Side A = 8.287023589

WORKSHEET PAGES/NOTES

The Final Obtuse Triangle

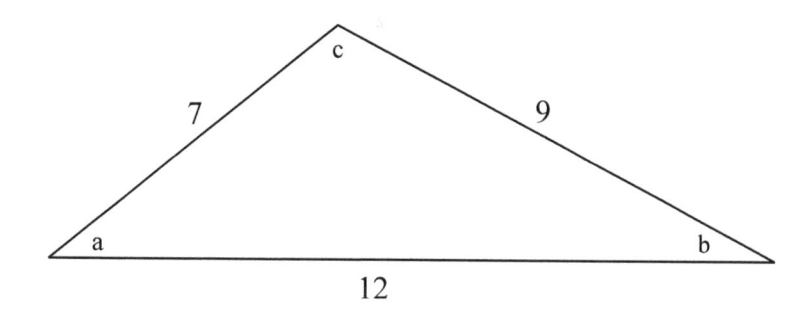

This is the final type of obtuse triangle. There are three **Sides** given but no angles. Once any of the angles are found, the **Law of Sines** can be used to finish off the rest of the triangle with ease.

This triangle uses the same formulas from the original **Law of Cosines**, but this time, we are looking for an **Angle** instead of a **Side**. The rules of how the formulas are computed are also different. The easiest way to go about learning this is to choose an angle to solve and go ahead and set up a formula from the **Law of Cosines** to match. Use **Angle a** as an example and insert the given information into the formula.

$9^2 = 7^2 + 12^2 - 2 \times 7 \times 12 \times$ **The Cosine of Angle a**

Like before, the problem can be divided into separate brackets for visual understanding. This time, however, there are four brackets of information instead of three.

Problem #1	Problem #2	Problem #3	Problem #4
$[9^2]$ =	$[7^2 + 12^2]$	$[-2 \times 7 \times 12]$ ×	[The Cosine of **Angle a**]

Multiplication symbol gets deleted

To make things easier, pretend that **Problem #4** is invisible. It will not be used or repositioned until all of the other steps have been done. Also, delete the multiplication symbol as noted above.

Solve the math in the first three problems and then rewrite the formula.

Now this is where it gets tricky. The solution from **Problem #1** will be manipulated and changed by the solution from **Problem #2** and **Problem #3** individually, and in a specified order, beginning with **Problem #2**.

[81] = [193] [-168]

Here are the rules:

If the answer from **Problem #2** is a **negative number**, it will be **ADDED** to the answer from **Problem #1**.
If the answer from **Problem #2** is a **positive number**, which it is in this case, it will be **SUBTRACTED** from the solution of **Problem #1**.
Once the addition or subtraction is completed, **Problem #2** gets deleted from the formula.

Subtract **Problem #2** from **Problem #1**. Then rewrite the formula again.

81 – 193 = -112

[-112] = [-168]   Continued on next page… ⟶

[-112] = [-168] ◄——— Carried over from previous page ...

Now that **Problem #2** has been subtracted from **Problem #1**, it's time to modify **Problem #1** again by using the last remaining solution from **Problem #3**.

The rules:
When this stage is reached, **Problem #1** is <u>**ALWAYS DIVIDED**</u> by **Problem #3**.
Once the division is completed. **Problem #3** gets deleted from the formula.

Divide the two remaining problems.
(Remember, a negative number divided by another negative number will result in a positive number)

-112 ÷ -168 = 0.666666667

Rewrite the formula. This time, insert **Problem #4** on the right side of the = symbol.

0.666666667 = The Cosine of Angle a

The final step is to convert the decimal back into an angle using **COS**$^{-1}$ with the calculator.

Angle a = 48.1896851

The **Law of Sines** could now be used to finish off the triangle, but for practice, take the long route and solve for **Angle b** using the **Law of Cosines**.

Below is a complete breakdown using the same steps from before to find **Angle b**.

Side B^2 = Side A^2 + Side C^2 - 2 × Side A × Side C × The Cosine of Angle b

[7^2] = [9^2 + 12^2] [- 2 × 9 × 12] × [The Cosine of Angle b]

49 = 225 -216
  Subtract

-176 = -216
  Divide

0.814814815 = The Cosine of Angle b

Angle b = 35.43094469

Find the final angle by subtracting the two solved angles from 180.

180 - 48.1896851 - 35.43094469 = 96.37937021

Angle c = 96.37937021

The triangle is now fully solved.

Practice Problems

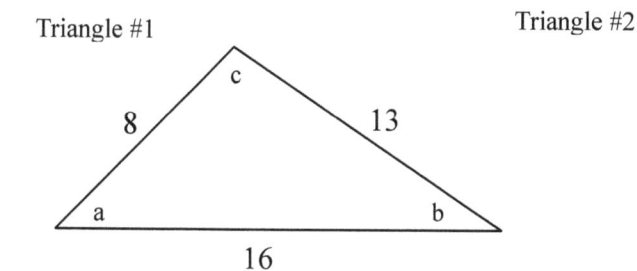

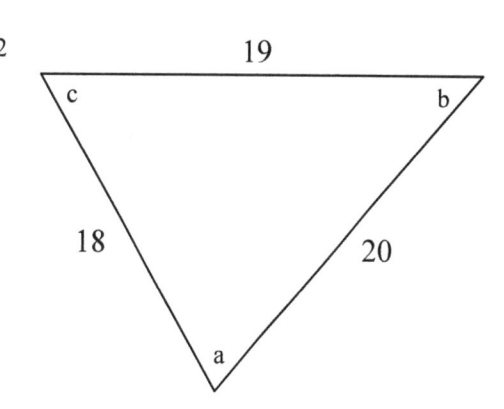

Triangle #1

Triangle #2

Answers:

Triangle #1 (Solving for **Angle a**)

Side A^2 = Side B^2 + Side C^2 - 2 × Side B × Side C × The Cosine of Angle a
13^2 = 8^2 + 16^2 - 2 × 8 × 16 × The Cosine of Angle a
169 = 320 -256
-151 = -256
0.58984375 = The Cosine of Angle a

Angle a = 53.85407899

Now solve for either of the two remaining angles. Use **Angle b**.

Side B^2 = Side A^2 + Side C^2 - 2 × Side A × Side C × The Cosine of Angle b
8^2 = 13^2 + 16^2 - 2 × 13 × 16 × The Cosine of Angle b
64 = 425 -416
-361 = -416
0.867788462 = The Cosine of Angle b

Angle b = 29.7973473

Now subtract the two angles from 180 to find **Angle c**.

180 – 53.85407899 – 29.7973473 = Angle c

Angle c = 96.34857371

The triangle is solved.

27

Triangle #2 (Solving for **Angle c**)

Side C^2 = Side A^2 + Side B^2 - 2 × Side A × Side B × The Cosine of Angle c
$20^2 = 19^2 + 18^2$ - 2 × 19 × 18 × The Cosine of Angle c
400 = 685 -684
-285 = -684
0.416666667 = The Cosine of Angle c

Angle c = 65.37568165

Solve for one of the remaining angles. Let's use **Angle b**.

Side B^2 = Side A^2 + Side C^2 - 2 × Side A × Side C × The Cosine of Angle b
$18^2 = 19^2 + 20^2$ - 2 × 19 × 20 × The Cosine of Angle b
324 = 761 -760
-437 = -760
0.575 = The Cosine of Angle b

Angle b = 54.9003678

Subtract the two angles from 180 to find **Angle a**.

180 – 65.37568165 – 54.9003678 = Angle a

Angle a = 59.72395055

The triangle is complete.

WORKSHEET PAGES/NOTES

Chapter 7
Miscellaneous Information

Isosceles Triangles

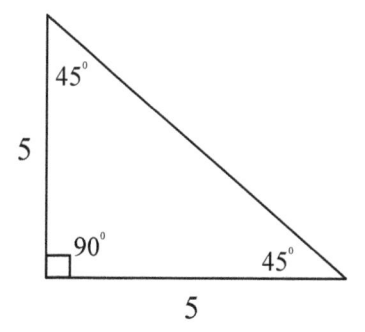

Any Right Triangle or Obtuse Triangle that has two **Legs** or two **Sides** that have the same dimension is called an **Isosceles Triangle**. In a Right Triangle, the two acute angles will always be 45 degrees. So, when faced with a Right Triangle containing a 45-degree angle, it can instantly be deducted that both **Legs** are the same dimension, and the other angle is also 45 degrees. This type is commonly known as a **45 – 45 – 90** triangle. It is a valuable shortcut when committed to memory.

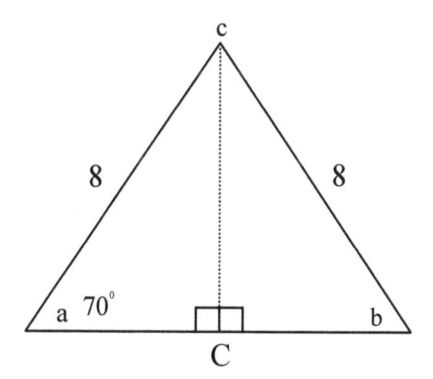

This special kind of **Obtuse Triangle** can be solved using Right Angle Trigonometry. If an imaginary line is drawn, starting from the center of the angle where the two given **Sides** intersect, to the middle of the **Opposite Side**, it creates two Right Triangles that are mirror images of each other. In the triangle above, **Angle a** and the **Hypotenuse** have been given. Using **Cosine** from Right Angle Trig, the **Adjacent Leg** to **Angle a** can be found.

Once this is done, the result can be multiplied by 2 in order to find the dimension for **Side C**. Since the triangles are mirror images, **Angle b** must also be 70 degrees. Now all that needs to be done is to subtract the two angles from 180 to find **Angle c**, and the triangle will be complete. This is useful if there is an Obtuse Triangle with two **Sides** that are the same and you do not want to use the **Law of Sines** or the **Law of Cosines** to solve it.

29

Rules for Creating Your Own Practice Problems

When creating problems to solve in trigonometry, you have to know a few things about triangles that were not covered earlier in the book.

Look at the two lines below and the triangle on the right.

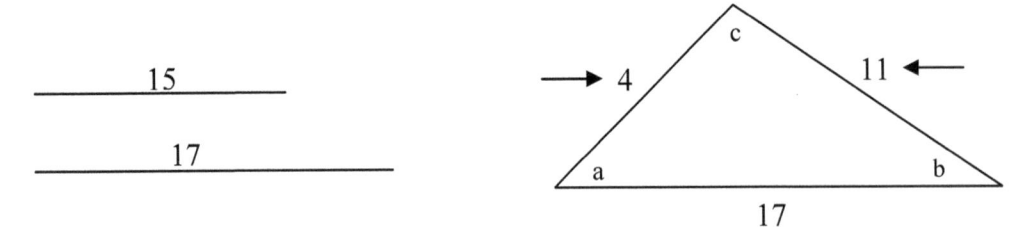

If you add up **11 + 4** from the sides of the triangle, the result is **15** which is the length of the smaller line above. **Side C** is **17**. The triangle is false because the smaller line would disappear into the longer one if they were positioned on top of each other. In reality, if an angle were applied to the line that is **15** in length, it would look something like this:

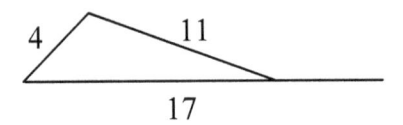

The line that is **17** in length extends past where the angle is formed by the intersection of the two lines.

There is a rule. **The length of any two Sides of a triangle must add up to be greater than the third Side**. The math must work out all the way around the whole triangle. No matter which two **Sides** are added together, the one left standing must be smaller than their sum or the triangle is false. This applies to both Right and Obtuse Triangles. However, in Right Triangles, the **Hypotenuse** must be larger than either of the two **Legs** because it is always the longest side. It is recommended, when setting up a Right Triangle with both **Legs** and the **Hypotenuse** given, that you choose the dimensions for one of the **Legs** and the **Hypotenuse** and then use the **Pythagorean Theorem** to get the real dimension for the last **Leg**. If it is done this way, the triangle will always be absolutely true. The same concept can be used when creating certain types of Obtuse Triangles as well. If an error is encountered while solving a triangle you have created, analyze it closely to see that it is a true triangle before becoming frustrated. You may be doing the right math, but if the triangle is false, it cannot be solved properly and may even result in an "ERROR" message on the display of the calculator.

Tips and Tricks

As more experience with Trigonometry is gleaned, intuition gives rise to certain techniques and methods that can make solving triangles even easier. These are the "Tricks" of the trade. One of the most popular tricks is to use Right Angle Trigonometry to solve certain types of Obtuse Triangles. The one below is a great example.

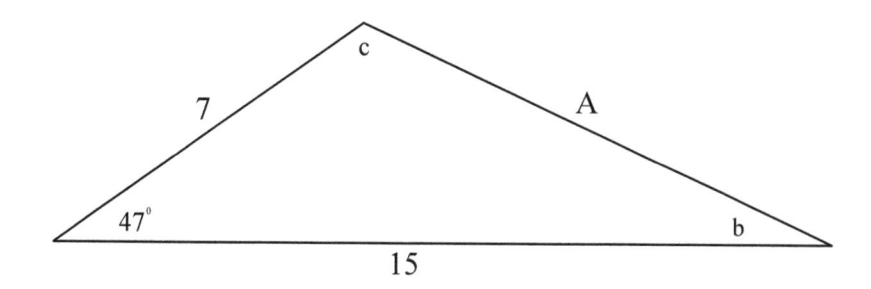

Normally, the **Law of Cosines** would be applied to solve this triangle. There is another way that is not only faster but also allows for the determination of another dimension that wasn't covered at all earlier in the book, the height of the triangle. If an imaginary line is drawn from the center of **Angle c**, straight down to intersect **Side C**, it forms two Right Triangles. The concept is very similar to the methods applied to Isosceles Triangles. The catch this time is that the two triangles are different sizes and have different angles.

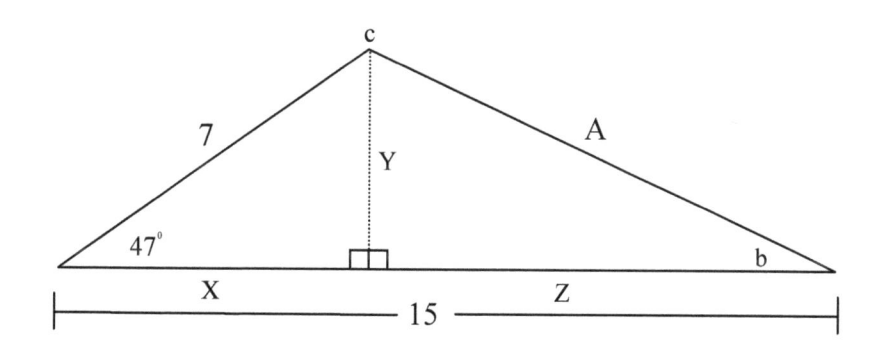

By looking at the picture, it is evident that the two triangles share the dimension of the newly drawn line labeled **Side Y**. **Side C** of the original Obtuse Triangle is divided into two new **Legs** labeled **Side X** and **Side Z**. If the dimension for **Side X** can be solved using the triangle on the left, then **Side Z** can be found by subtracting **Side X** from 15. Then, once **Side Y** is computed, enough information will be available to finish solving the triangle on the right, except for **Angle c**, which can be found by subtracting **Angle a** and **Angle b** from 180 or adding the two smaller angles from both Right Triangles together.

Answers:

Side X = 4.77398852
Side Y = 5.119475911 (Height of Triangle)
Side Z = 10.22601148
Side A = 11.43592342
Angle b = 26.59404545
Angle c = 106.4059546

Another trick is the method in which the scientific calculator is used. Often times, there will be a dimension in the display that carries out quite a few decimal places. This number may need to be used in conjunction with another dimension during a calculation. The best way to retain the number is to store it into the memory and then recall it later when needed. Another trick, when finding the **Difference** between two numbers, is to subtract the other number that you are using from the one in the display. For example:

57.26598741 is an angle that is displayed on the screen of the calculator. You are looking for the other angle in a Right Triangle. Instead of storing the number into the memory or typing it out again, simply subtract 90 from it.

(57.26598741) – 90 = -32.73401259

All that needs to be done now is to **disregard the negative symbol**. Subtracting, even when the result is a negative number, is the easiest way to find the difference between two dimensions. If the new number is needed for the next calculation, simply change it to a positive by using the conversion key on the calculator. It should be noted, however, that not all calculators support this conversion ability. An even better example applies when solving angles in an Obtuse Triangle.

33.54813659 is an angle that is displayed on the screen. It is the second angle of three in an Obtuse Triangle. The other angle given is **40 degrees**. You are solving for the final angle. Rather than typing 180 and then subtracting the angles in sequence, this way is a lot faster:

(33.54813659) + 40 – 180 = -106.4518634

The correct answer for the third angle is now on the display as a negative number but we know that the correct answer is a positive.

Congratulations! You now have most of the basics necessary to become a master at Trigonometry. With a little common sense and intuition, you will be able to form triangles out of all sorts of data and solve for missing information and dimensions across a wide variety of professions.

Whatever your application is, I wish you the best of luck in your endeavor. Thank you for choosing my book. I hope it will be a valuable addition to the tools of your success.

Sincerely

Kevin D Hunter
Author

Quick Reference

Tripod Structure & Definition

$$B = \frac{C}{A}$$

$A \times B = C$
$C \div B = A$
$C \div A = B$

Anatomy of Right Triangles

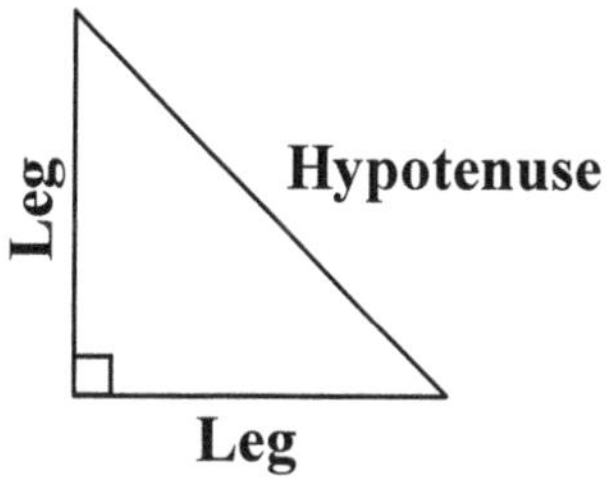

The Three Tripods of Right Angle Trigonometry

Sine $\text{Angle} = \frac{\text{Opposite Leg}}{\text{Hypotenuse}}$

Cosine $\text{Angle} = \frac{\text{Adjacent Leg}}{\text{Hypotenuse}}$

Tangent $\text{Angle} = \frac{\text{Opposite Leg}}{\text{Adjacent Leg}}$

The Pythagorean Theorem

$A^2 + B^2 = C^2$
$C^2 - B^2 = A^2$
$C^2 - A^2 = B^2$

Quick Reference

The Law of Sines

Side A ÷ the Sine of Angle a is the same as ⇩
Side B ÷ the Sine of Angle b, which is also the same as ⇩
Side C ÷ the Sine of Angle c

Also Known as:

(A/SIN a) = (B/SIN b) = (C/SIN c)

The Law of Cosines

To solve for **Side A** or **Angle a**
Side A^2 = Side B^2 + Side C^2 - 2 × Side B × Side C × The Cosine of Angle a

To solve for **Side B** or **Angle b**
Side B^2 = Side A^2 + Side C^2 - 2 × Side A × Side C × The Cosine of Angle b

To solve for **Side C** or **Angle c**
Side C^2 = Side A^2 + Side B^2 - 2 × Side A × Side B × The Cosine of Angle c

Useful Information

Acute Angle – Any angle less than 90 degrees but greater than zero

Obtuse Angle – Any angle greater than 90 degrees but less than 180

Inverse Angle – Angle calculated after subtracting original angle from 180. Used in cases where the **Law of Sines** produces an **Acute Angle** when the real angle is **greater than 90 degrees**

Adjacent – Directly next to or parallel to

Isosceles – Any three-sided polygon with two equal sides

Constant – A number or value that never changes

Variable – An unknown numeric value represented by letters, drawings or symbols